W9-DAB-333

INVESTIGATING
EARTH'S
POLAR
BIOMES

INVESTIGATING EARTH'S POLAR BIOMES

EDITED BY SHERMAN HOLLAR

Britannica®
Educational Publishing

IN ASSOCIATION WITH

ROSEN
EDUCATIONAL SERVICES

Published in 2012 by Britannica Educational Publishing
(a trademark of Encyclopædia Britannica, Inc.)
in association with Rosen Educational Services, LLC
29 East 21st Street, New York, NY 10010.

Distributed exclusively by Rosen Educational Services.
For a listing of additional Britannica Educational Publishing titles, call toll free (800) 237-9932.

First Edition

Britannica Educational Publishing
Michael I. Levy: Executive Editor, Encyclopædia Britannica
J.E. Luebering: Director, Core Reference Group, Encyclopædia Britannica
Adam Augustyn: Assistant Manager, Encyclopædia Britannica

Anthony L. Green: Editor, Compton's by Britannica
Michael Anderson: Senior Editor, Compton's by Britannica
Sherman Hollar: Associate Editor, Compton's by Britannica

Marilyn L. Barton: Senior Coordinator, Production Control
Steven Bosco: Director, Editorial Technologies
Lisa S. Braucher: Senior Producer and Data Editor
Yvette Charboneau: Senior Copy Editor
Kathy Nakamura: Manager, Media Acquisition

Rosen Educational Services
Alexandra Hanson-Harding: Editor
Nelson Sá: Art Director
Cindy Reiman: Photography Manager
Matthew Cauli: Designer
Introduction by Alexandra Hanson-Harding

Library of Congress Cataloging-in-Publication Data

Investigating earth's polar biomes / edited by Sherman Hollar. —1st ed.
    p. cm.—(Introduction to earth science)
"In association with Britannica Educational Publishing, Rosen Educational Services."
Includes bibliographical references and index.
ISBN 978-1-61530-501-8 (library binding)
1. Biotic communities—Polar regions—Juvenile literature. 2. Ecology—Polar regions—Juvenile
literature. I. Hollar, Sherman.
QH84.1.I585 2012
577.0911—dc22

                                                                              2010048951

Manufactured in the United States of America

On the cover, page 3: An iceberg is reflected in beautiful still water off the coast of Greenland.
Shutterstock.com

Interior background: © www.istockphoto.com/brytta

CONTENTS

INTRODUCTION   6

CHAPTER 1 GEOGRAPHY OF THE ARCTIC REGIONS   10

CHAPTER 2 ARCTIC EXPLORATION AND SCIENTIFIC RESEARCH   28

CHAPTER 3 GEOGRAPHY OF ANTARCTICA   43

CHAPTER 4 ANTARCTIC RESOURCES   60

CHAPTER 5 STUDY AND EXPLORATION OF ANTARCTICA   70

CONCLUSION   79
GLOSSARY   81
FOR MORE INFORMATION   83
BIBLIOGRAPHY   86
INDEX   87

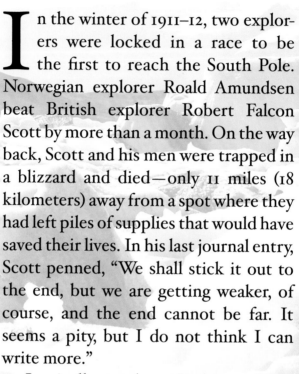

In the winter of 1911–12, two explorers were locked in a race to be the first to reach the South Pole. Norwegian explorer Roald Amundsen beat British explorer Robert Falcon Scott by more than a month. On the way back, Scott and his men were trapped in a blizzard and died—only 11 miles (18 kilometers) away from a spot where they had left piles of supplies that would have saved their lives. In his last journal entry, Scott penned, "We shall stick it out to the end, but we are getting weaker, of course, and the end cannot be far. It seems a pity, but I do not think I can write more."

Ironically, only 16 years later, Amundsen lost his life in a plane crash in the Arctic. These explorers faced frostbite, cruel whipping winds, danger, and even death to learn about these forbidding regions. But with this volume, you can learn all about the polar biomes—without the trouble of putting on snowshoes or even a warm jacket.

The first biome you will learn about is the North Pole (a biome is a major ecological community

Ecotourists maneuver between icebergs on a boat in Erera Channel, Antarctica. Keenpress/National Geographic Image Collection/Getty Images

type, such as a forest, grassland, or desert). The Arctic can be defined as the area within the Arctic Circle, which extends 1,650 miles (2,660 kilometers) from the geographic North Pole. Or it can be defined as the northern area in which the mean temperature for the warmest month is less than 50 °F (10 °C). The pole itself is covered with ice; under that ice is not land but the Arctic Ocean, though the amount and thickness of the ice varies from season to season.

But there are land areas that are considered part of the Arctic region as well—from Alaska and Canada to Russia and Scandinavia. These regions have special kinds of cold-weather ecosystems, such as boreal (northern) forests and large regions of permafrost, or permanently frozen ground. They are also home to a number of native peoples who have been "explorers" of the region in their own right, from the Inuit people who have traditionally hunted and fished in Greenland, Canada, northern Alaska, and Russia, to the Sami people who continue to herd reindeer in frigid parts of Scandinavia.

Unlike the Arctic, the other polar biome you will explore is a continent. But only two percent of Antarctica's land can be seen by the naked eye. The rest is covered by a thick ice sheet, which at one point is almost 3 miles (5 kilometers) deep. It contains 90 percent of the world's ice and 70 percent of the world's fresh water. Because it is on land, not water, Antarctica is even colder than the Arctic region. The world's record low temperature of –128.6 °F (–89.2 °C) was recorded there. The coast can be extremely windy, with gusts of nearly 200 miles (320 kilometers) per hour. Antarctica's interior is one of the world's major cold deserts. Precipitation averages only 1 to 2 inches (2.5 to 5 centimeters) a year.

Little survives in Antarctica's interior, except for a few tiny creatures such as mites. But the ocean that surrounds it is teeming with life, including krill, which feed whales, seals, and penguins. Antarctic cod have blood that lets them live in seawater as cold as 28 °F (–2 °C).

People still have the desire to conquer and understand new frontiers. Every summer, about 25 nations send approximately four thousand scientists to Antarctica to conduct research, including studying the continent's weather, climate, and mineral resources. As you read this volume, you too will come to understand why people are still fascinated by these icy regions.

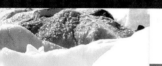

CHAPTER 1
GEOGRAPHY OF THE ARCTIC REGIONS

A vital zone between North America's and Russia's northernmost frontiers consists of the Arctic regions.

Once only explorers, traders, and Inuit, or Eskimo, hunters were interested in the vast, icy area at the "top" of the world. Today, because of its strategic location and its value to scientists, the Arctic is the scene of much activity. Year-round scientific research stations are maintained to study weather, climate, and mineral resources of the Arctic. In addition, the Arctic is studded with air bases, constant reminders that the shortest air routes between the United States and Russia are over the area. Only a narrow channel separates Little Diomede Island, of the United States, from Big Diomede Island (Ostrov Ratmanova), which is Russian territory. With the technological advances in icebreaker ships and nuclear-powered submarines, the distances between the two countries seem even shorter.

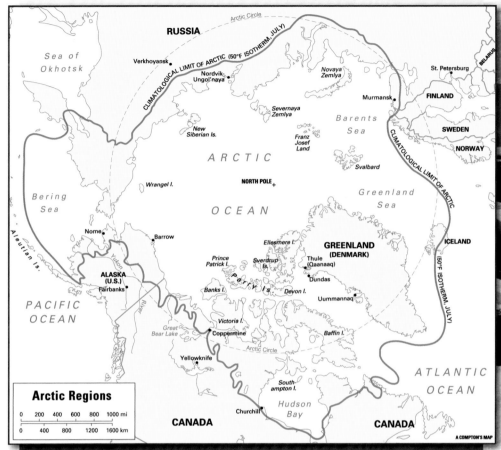

The average location of the 50 °F (10 °C) July isotherm in the Northern Hemisphere—the climatological limit of the Arctic. An isotherm is a line drawn on a map or chart joining points with the same temperature.

The Arctic is sometimes defined as the area that lies within the Arctic Circle. The Arctic Circle is a parallel of latitude (66°30' N. latitude), 1,650 miles (2,660 kilometers)

from the North Pole, the northern end of the Earth's axis. Actually, the Arctic Circle does not enclose all the Arctic regions. The true Arctic is the area in which the mean temperature for the warmest month is less than 50° F (10 °C). The coldest region, the "polar segment," is where the mean temperature of the warmest month is below freezing.

The subarctic region is the area that has a mean temperature above 50 °F (10 °C) for more than three but less than four months a year. The boundary of the Arctic is sometimes said to be the line beyond which no trees grow. This is based on the theory that tree life cannot exist unless there is at least one month a year with a temperature of 50 °F (10 °C).

ARCTIC OCEAN AND ARCTIC LAND

The greater part of the 8 million square miles (21 million square kilometers) within the Arctic Circle is occupied by the Arctic Ocean (5,440,000 square miles, or 14,090,000 square kilometers). Around the pole, the ocean is about 13,800 feet deep (4,200 meters). Islands dot the southern two thirds

Colorful houses and landscape in Greenland. **iStockphoto/Thinkstock**

of the ocean. Below these comes a rim of land provided by the northern continents.

The most important islands north of America are Baffin, Victoria, and Ellesmere, belonging to Canada. Svalbard, a Norwegian group, and Franz Josef Land and Novaya Zemlya, belonging to Russia, are the largest islands north of Europe.

North of Asia, near Siberia, lie Severnaya Zemlya and the New Siberian Islands. Other

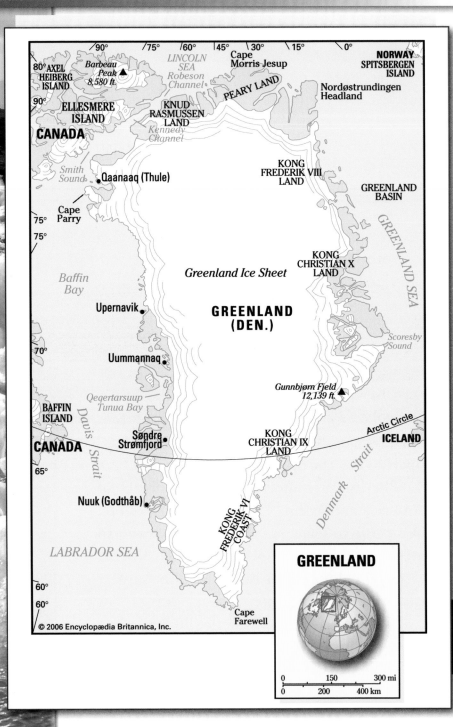

islands in the Arctic regions are Wrangel, Prince Patrick, Devon, and Banks, as well as the Parry Islands. Alaska and northern Canada form the Arctic lands of North America. Farther east is the world's largest island, Greenland. It is part of the Danish kingdom. At Dundas, Greenland, the United States has a large air base.

In northern Europe the Laplands of Norway, Sweden, Finland, and western Russia jut into the Arctic. Eastward stretches the Russian territory of Siberia.

In much of the Arctic, earth, ice, and rock are frozen solid permanently. The solid mass is called permafrost. It is covered with a layer of ice and snow that melts in summer. In winter the Earth's crust in the Arctic is a solid frozen mass as deep as 1,000 feet (300 meters) in some places. Climate warming in recent years, however, has caused part of the upper permafrost layer to melt. The result in summer is a deep layer of swampy land, which can cause buildings to sink. In

Map of Greenland highlighting the major geographic regions and the locations of human settlement.

A so-called drunken forest of black spruce in Fairbanks, Alaska. A drunken forest is one where trees that originally grew straight and tall in permafrost start to tilt if the permafrost starts to melt, leaving the soil too loose to hold their roots firmly. **Ashley Cooper/Visuals Unlimited/Getty Images**

addition, stored carbon gases may be released from the once-frozen soil, contributing to global warming.

CLIMATE

The most extreme winter cold and summer heat in the Arctic are not at the pole because the Arctic Ocean prevents extremes. The water absorbs heat during the summer and gives it out in the winter.

Greater extremes occur near the Arctic Circle because the land there is less effective than water in storing heat. Alaska has had a winter temperature of −80°F (−62°C). In summer the temperature has reached 100°F (38°C). The coldest weather in the Arctic regions occurs near Verkhoyansk in Siberia. The January temperature there can reach −90°F (−68°C). The Arctic is warmer than Antarctica.

Within the Arctic Circle winter cold is bearable because there is little wind. Blizzards and gales occur only when the air is flowing strongly outward across the Arctic Circle or where a break in the land level disturbs the circulation. The winter air is dry. Most of the moisture in the region is frozen.

The climate of the Arctic biome is determined by the amount of heat and light received from the sun. The slant of the Earth in relation to the sun prevents the sun's rays from reaching the Arctic regions for part of each year. The North Pole has no direct sunlight for six months. In summer the Arctic has long hours of sunlight. The sun's rays

Ice floes in the Arctic Ocean. **Photos.com/Thinkstock**

strike at a great slant, however, and do not give as much heat as they do farther south.

More than half the Arctic Ocean is covered with a layer of ice all the time. Much of it stays in place as a jumbled mass called pack ice.

However, in recent years, there has been a drastic thinning and reduction in the Arctic's pack ice. Many scientists cite global warming as a major factor. The leading scientific international organization in the study of global warming, the Intergovernmental Panel on Climate Change (IPCC), has projected that the Arctic could be virtually free of summer sea ice by 2070, and more recent studies project this could occur several decades sooner.

PLANT AND ANIMAL LIFE

There are many living things in the Arctic biome. Although not as numerous as in other oceans, countless microscopic plants called diatoms live in the polar sea. They furnish food for shrimps and other crustaceans. These, in turn, are eaten by fishes. Both the crustaceans and the fishes are eaten by seals, walruses, and the few whales that still live in the sea.

Much of the land in the Arctic is covered with a treeless grassy carpet called tundra.

ENVIRONMENTAL THREATS

Increased economic activity in the Arctic Ocean has caused considerable environmental concern. Habitats and living patterns of wildlife and sea life have been disturbed. The potential dangers from oil spills and other forms of pollution are immense. Because of the harsh cold, many areas that border the pack ice are open for shipping and construction of drilling platforms for only a few weeks each year. When this "summer" season is over, the ocean freezes and becomes impassable. This raises the danger that an oil leak might continue for months before it could be stopped and the spill cleaned up. Also, in the frigid Arctic climate, the rate at which crude petroleum and other pollutants decompose into environmentally harmless components is extremely slow.

Global warming, by reducing the Arctic Ocean's ice cover, has become a threat to the world's well-being, according to many scientists. Melting pack ice could raise sea levels throughout the world. In addition, Arctic animals are slowly dwindling as a result of habitat loss, threatening the traditional lifestyle of indigenous peoples.

So far the Arctic Ocean has survived increased human activity and environmental interference. Scientists are seeking ways to exploit its abundant resources without damaging the fragile environment.

In summer flowers and grasses spring up in some places. Caribou, reindeer, and musk oxen eat the tundra. In winter they paw away the snow to get moss and lichens.

Polar bears and wolves prey on these larger animals and the sea animals. Smaller meat eaters live mainly on sea life.

During the long cold season fox, ermine, muskrat, beaver, marten, mink, and other

A herd of reindeer grazes for food in the Arctic winter. Hemera/ Thinkstock

animals grow thick, rich furs. In the lengthy days of summer many species of migrating land birds, such as the redpoll, snowbird, pipit, and rock ptarmigan, fatten on the swarming insects. Seabirds eat the abundant fish. The snowy owl and the raven often winter here.

POLAR BEARS

The polar bear *(Ursus maritimus)* is a large and potentially dangerous bear. Polar bears are stocky, with a long neck, relatively small head, short, rounded ears, and a short tail. The male, which is much larger than the female, weighs 900 to 1,600 pounds (410 to 720 kilograms). It grows to about 5.3 feet (1.6 meters) tall at the shoulder and 7.2–8.2 feet (2.2–2.5 meters) in length. The tail is 3–5 inches (7–12 centimeters) long. Sunlight can pass through the thick fur, its heat being absorbed by the bear's black skin. Under the skin is a layer of insulating fat. The broad feet have hairy soles to protect and insulate as well as to facilitate movement across ice, as does the uneven skin on the soles of the feet, which helps to prevent slipping. Strong, sharp claws are also important for gaining traction, for digging through ice, and for killing prey. The most carnivorous of bears, it lives on ice far from land and on coastal areas and islands of the Arctic Ocean.

Adult polar bears have no natural predators, though walruses and wolves can kill them. Longevity in the wild is 25 to 30 years, but in captivity several polar bears have lived to more than 35 years old.

Humans probably cause most polar bear deaths by hunting and by destruction of problem animals near settlements. Polar bears have been known to kill people. The bears are hunted especially by Inuit people for their hides, tendons, fat, and flesh. Although polar bear meat is consumed by aboriginals, the liver is inedible and often poisonous because of its high vitamin A content.

A polar bear out for a walk on snow-covered tundra.
Shutterstock.com

At the turn of the 21st century, an estimated 20,000 to 25,000 polar bears existed in the wild. Because of continued global warming, a substantial reduction in the coverage of Arctic summer sea ice—prime habitat for polar bears—is expected by the middle of the 21st century. Models developed by some scientists predict an increase in polar bear starvation as a result of longer ice-free seasons and a decline in mating success, since sea-ice fragmentation could reduce encounter rates between males and females. Model forecasts by the U.S. Geological Survey suggest that habitat loss may cause polar bear populations to decline by two-thirds by the year 2050. In May 2008 the U.S. government listed the polar bear as a threatened species.

PEOPLE OF THE ARCTIC

The Arctic plains have only a sparse and scattered native population. Traditionally the Inuit who dwelled in Arctic America and Greenland depended upon hunting and fishing. Their permanent homes were made of sod, stones, and driftwood, and they used igloos or tents of animal skins for temporary shelter. In the 20th century, however, the Inuit became increasingly influenced by societies to the south. Modern manufactured goods such as snowmobiles, rifles, and store-bought clothing have entered the culture. Many Inuit have abandoned their nomadic hunting pursuits to move into towns or to work in mines and oil fields. Others, particularly in Canada, have formed cooperatives to market their handicrafts, fish catches, and ventures in tourism.

Until the 21st century the Sami (Lapps) of northern Europe roamed about the tundra with their grazing herds of reindeer. In the Asian Arctic, other reindeer-herding tribes live in Siberia. Other people of the Asian Arctic hunt and fish in summer and trap fur-bearing animals in winter.

LAPLAND

The region called Lapland stretches across Arctic Norway, Sweden, and Finland and includes the Kola Peninsula of Russia. It is bounded by the Norwegian Sea on the west and the Arctic Ocean on the north and east. The western part of Lapland contains high mountains that are deeply eroded into fjords and headlands in Norway. Across the border, in Sweden, Lapland contains that country's highest peaks. From these peaks the land

slopes downward to the east where in Finland and Russia it becomes low-lying and marshy tundra. All of Lapland is windy, but sheltered swamplands and river valleys support natural meadowland. Game birds are abundant and waterways are well stocked with fish.

Native Sami (Lapps) are citizens of the country in which they maintain permanent villages. The origin of the Sami is uncertain—they may be descendants of one of the original Finnic tribes in the Baltic region or they may be descendants of immigrant Siberian tribes. In either case they were present north of the Gulf of Finland long before any of the other present-day Baltic peoples. The three Sami languages belong to the Finno-Ugric group and are related to the languages of the Finns and Hungarians. A speaker of one Sami language normally cannot understand the other two Sami languages, though the three sometimes are considered dialects of the same language.

Nomadic reindeer herding was the traditional way of life for most Sami until the late 20th century. Although the wild reindeer herds were largely domesticated, the people went wherever the animals can find lichens (reindeer moss) to eat. Reindeer provided milk, cheese, and meat for food, and skins for tents, blankets, moccasins, leggings, and harnesses. The nomadic Sami lived in tents made from skins stretched over poles. Reindeer-drawn sleds, called pulkas, provided transportation over frozen areas. The traditional Sami

dress is a blue pullover tunic decorated with red and yellow trim, with a fringed shawl and a red and blue bonnet for women and blue breeches and a large four-cornered blue cap with red pom-poms for men.

In the 21st century, few Sami remain wholly nomadic. The herders generally travel with their reindeer alone; their families stay behind year-round in modern housing. The Sami who live along the coasts are engaged largely in fishing, farming, and livestock raising. Because of the short growing season, agriculture is limited. The chief crops are potatoes, barley, and rye. Sami also work in forestry, industry, mining, government, and commerce. They increasingly have been integrated into the larger Scandinavian workforce and culture. Important towns in Lapland include Tornio, Kemi, and Luleå, all seaports on the Gulf of Bothnia.

A traditionally dressed Sami herdswoman, with her reindeer, in Norway. Jorn Georg Tomter/ Stone+/Getty Images

CHAPTER 2

ARCTIC EXPLORATION AND SCIENTIFIC RESEARCH

More than a hundred years ago, people became interested in exploring the Arctic for scientific purposes. Geologists knew that an Ice Age had once engulfed much of Europe and North America. They hoped to learn more about it by studying Arctic ice. Geographers were also interested in the Arctic itself. They wanted to find out why the magnetic north pole was not at the geographic pole.

SCIENTIFIC INTEREST IN THE ARCTIC

These and other scientific motives led explorers to make expeditions by ship and then across the ice by dog sled. Many efforts ended in tragedy, however. In 1845 Sir John Franklin, with 128 other explorers, tried to sail the Northwest Passage in two ships, the *Erebus* and the *Terror*. For several years nothing was heard of the men. Then, in 1859, a rescue party found their bodies on King

William Island near Victoria Strait, where the ice had trapped their ships. With them was a record. Its last date was April 25, 1848.

Another tragedy resulted from an international attempt at polar research. Representatives from 10 nations met in Hamburg, Germany, in 1879, and in Bern, Switzerland, in 1880. They decided to set up scientific stations in the far north. The United States agreed to operate bases at Point Barrow, Alaska, and at Lady Franklin Bay, on Ellesmere Island. In 1881 an expedition headed by A. W. Greely began work on Ellesmere Island, only 497 miles (800 kilometers) from the North Pole. The party gathered valuable information. Two relief expeditions, however, failed to reach them. Sixteen of the 23 men died of starvation on Cape Sabine before help finally came.

CONQUERING THE "PASSAGES"

Not all the early attempts at Arctic exploration ended in tragedy or failure. In 1850 Capt. Robert McClure in the ship *Investigator* headed east from the Bering Sea in search of the John Franklin expedition. His ship was frozen into the ice during three winters. In the spring of 1853 McClure abandoned it. He led his crew eastward over the ice to Melville Island. There a rescue ship was waiting for them. Thus they traversed the Northwest Passage.

Twenty-five years later a Swedish explorer and scientist, Nils Adolf Erik Nordenskjöld, forced his way through the Northeast Passage. He left Tromsö, Norway, in the ship *Vega* in July 1878. He reached Yokohama, Japan, one year and two months later.

Early in the 20th century Roald Amundsen of Norway traversed the Northwest Passage. He sailed west from Greenland in the summer of 1905. His ship was frozen in the ice during the winter. The following spring he worked it free and pushed through to the Bering Sea and the Pacific.

Roald Amundsen

One of the most important men in the history of polar exploration was Roald Amundsen. He was the first man to reach the South Pole, the first to sail around the world via the Northwest and Northeast passages, and the first to fly over the North Pole in a dirigible.

Amundsen was born in Borge, Norway, on July 16, 1872. His father, a shipowner, died when the boy was 14. In school young Amundsen read stories about Sir John Franklin and other polar explorers and set his heart on becoming an explorer himself.

At the age of 25 Amundsen became the first mate of the ship *Belgica* on a Belgian expedition to the Antarctic. After he returned to Norway, he prepared for his first independent venture.

In 1903 Amundsen set sail in the ship *Gjöa*, hoping to locate the magnetic North Pole. For 19 months he remained in King Wilhelm Land, in the northeastern part of Greenland, making observations. His studies indicated that the magnetic pole has no stationary position but is in continual movement.

Norwegian polar explorer Roald Amundsen on his arrival in Alaska after a transpolar flight in 1926. Topical Press Agency/ Hulton Archive/Getty Images

While on this expedition, he also traversed, in 1905, what explorers had been seeking for more than three hundred years — the Northwest Passage from the Atlantic to the Pacific. He sailed through bays, straits, and sounds to the north of Canada. Thus Amundsen justified the search for a shorter route to the Orient that had challenged countless navigators.

Amundsen had planned next to drift across the North Pole in Fridtjof Nansen's ship *Fram*. When he learned that the American explorer Robert E. Peary had reached the North Pole in April 1909, he decided to seek the South Pole instead. He arrived there on Dec. 14, 1911 — just 35 days before the arrival of Robert F. Scott.

In the summer of 1918 Amundsen once again set sail for the Arctic, in the newly built ship *Maud*. He planned to drift across the North Pole from Asia to North America, but he failed in this purpose because the ship was unable to penetrate the polar ice pack. Two years later, however, when Amundsen reached Alaska, he had sailed also through the Northeast Passage, by way of Siberian coastal waters connecting the Atlantic with the Pacific.

After he had sent the *Maud* back to the Arctic to continue observations, Amundsen turned to the project of flying over the North Pole. His efforts were crowned with success on May 11–13, 1926. In the dirigible *Norge*, piloted by Col. Umberto Nobile, an Italian aviator, he flew over the pole on a 2,700-mile

(4,300-kilometer) flight between Spitsbergen and Teller, Alaska. He was accompanied by Lincoln Ellsworth, an American explorer.

Two years later Amundsen embarked on his last adventure. In June 1928 he left Norway to fly to the aid of Nobile, whose dirigible had crashed on a second Arctic flight. Amundsen's plane vanished, though Nobile was later rescued. Months afterward the discovery of floating wreckage told the tragic story of how Amundsen met his death.

THE QUEST FOR THE NORTH POLE

Meanwhile many explorers tried to reach the North Pole. A Norwegian scientist, Fridtjof Nansen, noticed that objects such as driftwood and a pair of trousers from a wrecked steamer were carried by floating ice across the Arctic Ocean from Asia to the North Atlantic. In 1893 he and 13 companions let their ship *Fram* be frozen into the ice pack north of Siberia.

Then for three years they slowly drifted westward. The drift, however, carried them wide of the North Pole. Nansen, with one companion, left the ship and dashed across the ice with a dog sled. They were unable to get farther north than 86°14'. In 1896 the

Norwegian explorer Fridtjof Nansen and his crew prepare to set sail in the Fram *for the North Pole in 1893.* Hulton Archive/Getty Images

ship finally came out of the ice pack north of Europe.

In 1897 Dr. Salomon Andrée of Sweden and two companions tried to drift over the pole in a balloon. Nothing was heard of them for 30 years. Then a Norwegian explorer, Dr. Gunnar Horn, found a photograph that had been taken by one of the party. It showed the balloon wrecked on White Island.

Robert E. Peary was the first man to reach the North Pole. He had spent 18 years in earlier

American explorer Robert Edwin Peary poses with his sled dogs in 1910. Imagno/Hulton Archive/Getty Images

Arctic exploration. Finally, on his eighth trip, he reached the pole on April 6, 1909.

Before the world received the news of Peary's victory, the following September, another American, Dr. F. A. Cook, arrived in Copenhagen, Denmark, claiming that he had reached the North Pole from Greenland on April 12, 1908. Peary and others challenged Cook's story, and it was later discredited. The test of being at the pole consists of seeing the sun and stars going around the sky in horizontal circles. Peary had 32 observations that met this test. Evidence was found suggesting that Cook's record of observations had been made up before he sailed.

Two other 20th-century explorers devoted their lives to studying the nature of the Arctic. They were Donald B. MacMillan, an American scientist, and Vilhjalmur Stefansson, a Canadian ethnologist. Stefansson wrote many books about his experiences in the "friendly Arctic." His writing helped convince the world that transarctic airlines were possible.

The New York Times *front page exposes explorer Frederick Cook's fraudulent claims to have reached the North Pole ahead of Robert E. Peary.* **Hulton Archive/Getty Images**

"All the News That's Fit to Print."

The New York Times.

THE WEATHER.
Warmer, increasing cloudiness to-day; showers to-morrow.

VOL. LVIII...NO. 18,856.   NEW YORK, THURSDAY, SEPTEMBER 9, 1909.—TWENTY PAGES.   ONE CENT

COOK NOT NEAR POLE, SAYS PEARY; PROOFS STILL HELD BACK BY COOK

Two Eskimos Cited by Cook Told Peary He Never Went Far from Land.

SAYS "I HAVE COOK NAILED"

Cables His Wife Not to Worry About the Doctor's Claim to the Pole.

COOK TO ASK FOR AN INQUIRY

Says He Will Submit His Observations to Competent Scientists.

DOUBT OF HIS STORY GROWS

Cook's Supporters Here and Through-out Europe Greatly Worried by His Failure to Refute Peary.

This message from Commander Peary to The New York Times was received at 4:57 o'clock yesterday morning:

INDIAN HARBOR, via Cape Ray, N. F., Sept. 8.—Cook's story should not be taken too seriously. The Eskimos who accompanied him say that he went no distance north. He did not get out of sight of land. Other men of the tribe corroborate their statements. Kindly give this to all home and foreign news associations for the same wide distribution as Cook's story.   PEARY.

Commander Peary also sent this message to The New York Times:

"Indian Harbor, Labrador.
"Via Cape Ray, N. F., Sept. 8
"Good morning. Delayed by gale. Don't let Cook story worry you. Have him nailed.
"BERT."

COOK DEMANDS A TRIBUNAL.

Still Declares He Has Proofs—With-holds Them, Despite Urging.

COPENHAGEN, Sept. 8.—This is Dr. Cook's reply to Commander Peary:

"I have been to the north pole. As I said last night when I heard of Com-mander Peary's success, if he says he has been to the north pole, I believe him.

"I am willing to place facts figures, and work-out observations before a joint tribunal of the scientific bodies of the world. In due course I shall be prepared to make public an an-nouncement that will effectually dispel any doubt, if there can be such, of the fact that I reached the pole. But, knowing that I am right and that this must prevail, I will submit at the proper time my full story to the court of last resort—the people of the world.

PROOFS TO-DAY, SAYS COOK.

But Grave Doubt is Felt That He Can Supply Them.

Special Cable to THE NEW YORK TIMES.

COPENHAGEN, Sept. 8.—The...

COMMANDER PEARY'S PRELIMINARY ACCOUNT OF HIS SUCCESSFUL VOYAGE TO THE NORTH POLE

He Sends to The Times by Wireless a Summary, to be Followed by His Full Report—Record of His Swift Progress to the Utmost North.

FROM CAPE COLUMBIA UP IN 37 DAYS, BACK IN 16 DAYS

Prof. Ross G. Marvin, of Cornell, Drowned on April 10, Forty-five Miles North of Cape Columbia, While Leading the Supporting Party.

BATTLE HARBOR, Labrador, Via Wireless Cape Ray, N. F., Sept. 8.—As it may be impossible to get my full story through in time for to-morrow's TIMES, partly as a prelude which may stimulate interest and partly to forestall possible leaks, I am sending you a brief summary of my voyage to the North Pole, which is to be printed exactly as written.

SUMMARY OF NORTH POLAR EXPEDITION OF THE PEARY ARCTIC CLUB.

The steamer Roosevelt left New York on July 6, 1908; left Sydney on July 17; arrived at Cape York, Greenland, August 1; left Etah, Greenland, August 8; arrived Cape Sheridan, at Granuland, September 1; wintered at Cape Sheridan.

The sledge expedition left the Roosevelt February 15, 1909, and started for the North. Arrived at Cape Columbia March 1; passed British record March 2; delayed by open water March 2 and 3; held up by open water March 4 to 11; crossed the 84th parallel March 11; encountered open lead March 15; crossed 85th parallel March 18; crossed 86th parallel March 23d; encountered open lead March 23d; passed Norwegian record March 23d; passed Italian record March 24th; encountered open lead March 26th; crossed 87th parallel March 27th; passed American record March 28; encountered open lead March 28; held up by open water March 29; crossed 88th parallel April 2; crossed 89th parallel April 4; North Pole April 6.

All returning left North Pole April 7, reached Cape Columbia April 23, arriving on board Roosevelt April 27.

The Roosevelt left Cape Sheridan July 18, passed Cape Sabine August 8; left Cape York August 26; arrived at Indian Harbor with all members of expedition returning in good health except Prof. Ross G. Marvin, unfortunately drowned April 10, when forty-five miles north of Cape Columbia, returning from 86° North Latitude in command of the supporting party.

ROBERT E. PEARY.

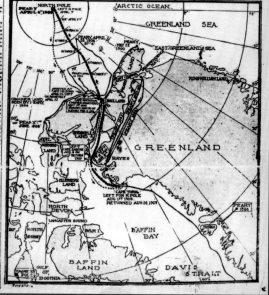

MAP SHOWING PEARY'S ROUTE TO THE POLE.

THE INTERNATIONAL GEOPHYSICAL YEAR (IGY)

The International Geophysical Year (IGY) was a worldwide program of geophysical research that was conducted from July 1957 to December 1958. The IGY was directed toward a systematic study of the Earth and its planetary environment. The IGY encompassed research in 11 fields of geophysics, from glaciology, gravity, longitude and latitude determinations, to meteorology. During the IGY, scientists studied weather and ice in the Arctic. Teams stationed on floes, or ice islands, found that their chief problem was summer thawing. Giant cracks appeared in one island, and the men and equipment had to be airlifted to another island.

Four such drifting ice stations (two Soviet, two American) were set up in the Arctic during the IGY. Driven by wind and current, they traveled as far as 4,000 miles (6,400 kilometers). Soundings made daily from one of the American stations revealed the existence of an unsuspected submerged mountain ridge. Another floe team collected samples of sediment and measured ocean depths, the Earth's magnetic field, temperatures, and ice movements. Studies made on McCall Glacier indicate that glaciers swell rather than shrink as the weather becomes warmer.

POST-IGY RESEARCH

The United States Navy's nuclear-powered submarines *Sargo* and *Sea Dragon* reached the North Pole in 1960. An ice island 12 feet (4 meters) high was found about 150 miles (240 kilometers) north of Point Barrow, Alaska, in May 1961. This floe became the United States Navy's Arctic Research Laboratory Ice Station II (ARLIS II). Weather and oceanographic studies were conducted there.

The United States Army Corps of Engineers has studied the properties of ice and problems of construction and living in the Arctic in two subsurface camps on the Greenland ice cap. At Camp Century, 138 miles (222 kilometers) from Thule Air Base on the western coast, engineers built tunnels and insulated buildings beneath the snow for their activities. At Camp Tuto tunnels were constructed through solid ice. Both camps have since been closed, though long- and short-term scientific studies are still carried out in the region.

In addition to an extensive scientific research program, Canada has encouraged

a widespread search for significant mineral deposits in its far north. It holds 28 percent of the Earth's Arctic land surface. Only Russia has more—40 percent. In the late 1960s huge reserves of natural gas were found on Melville and other Canadian Arctic islands, and in 1970 oil was discovered near the mouth of the Mackenzie River. Oil and gas were found in the shallow waters of the Beaufort Sea.

An aerial view of Prudhoe Bay, Alaska. Visible are the frozen Beaufort Sea, pack ice, and an oil rig. **Rich Reid/National Geographic Image Collection/Getty Images**

Extensive reserves of oil were discovered at Prudhoe Bay on Alaska's North Slope in 1968. Plans to build a pipeline to retrieve natural gas from the Mackenzie River area in the Arctic are ongoing.

The Arctic lands of Russia are known to contain rich reserves of minerals, including

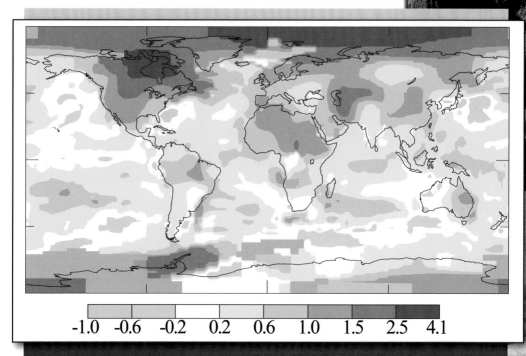

-1.0   -0.6   -0.2   0.2   0.6   1.0   1.5   2.5   4.1

The map shows in degrees Celsius the difference between the Earth's average annual temperature in 2006 (October 2005 to September 2006) and the average annual temperature during the base period 1951–80. The difference was greatest in the Arctic, where the annual average temperature had increased by as much as 4.1 °C (7.4 °F). **Encyclopædia Britannica, Inc.**

cobalt, nickel, coal, and iron. The gold and diamond mines of eastern Siberia place Russia high among world producers of these minerals. Natural gas and oil have been found in several places. The first nuclear power station in the Arctic was opened by the Soviet Union on the Kola Peninsula in 1974. In 1977 the Soviet nuclear icebreaker *Arktika* became the first ship to break its way through to the North Pole. Recent focus has been centered on discovering ways to clean up the Arctic, which was contaminated by Soviet nuclear and chemical testing before the end of the Cold War.

The purpose of the International Polar Year, taking place from March 2007 until March 2009, was to conduct further experiments in both the Arctic and Antarctic. A main goal of the initiative was to explore the effects of global warming, the results of which are magnified at the poles.

CHAPTER 3
GEOGRAPHY OF ANTARCTICA

The icy continent surrounding the South Pole is called Antarctica. This region is larger in area than Europe. It is a cold and forbidding land that has no permanent human population and is almost devoid of animal or plant life. However, the oceans adjoining Antarctica teem with life.

Ice and stormy seas kept anyone from seeing Antarctica until about 1820. In 1950 more than half the continent still had not been seen. Now airplanes and tractors have taken people to most parts of Antarctica, and satellite photographs have revealed the rest. But Antarctica remains a frontier, and much is yet to be learned about it.

Almost no one goes to Antarctica except scientists and some adventurous tourists. The continent has natural resources that someday may be used, but the harsh environment of the area makes them difficult to exploit. Nations interested in Antarctica have signed a treaty that reserves the region for science and other peaceful purposes.

THE LAND

An ice sheet covers nearly all of Antarctica. At its thickest point the ice sheet is 15,670 feet (4,776 meters) deep—almost 3 miles (5 kilometers). It averages 7,000 to 8,000 feet (2,100 to 2,400 meters) thick, making Antarctica the continent with the highest mean elevation. This ice sheet contains 90 percent of the world's ice and 70 percent of the world's fresh water.

The Antarctic ice was formed from the snows of millions of years that fell on the land, layer on layer. The weight of new snow squeezes the old snow underneath until it turns to a substance called firn, then ice. As the ice piles up, it moves toward the coast like batter spreading in a pan. The moving ice forms into glaciers, rivers of ice that flow into the sea. Pieces of the floating glaciers break off from time to time, a process called calving.

These icebergs float north until they reach warm water, break into pieces, and melt. Icebergs as large as 40 by 30 miles (64 by 48 kilometers) have been sighted, but most are smaller. In some places the floating glaciers stay attached to the land and continue to

Only a small part of a giant iceberg shows above the surface of the ocean. Shutterstock.com

grow until they become ice shelves. The Ross Ice Shelf alone is about the size of Canada's Yukon Territory and averages 1,100 feet (330 meters) thick.

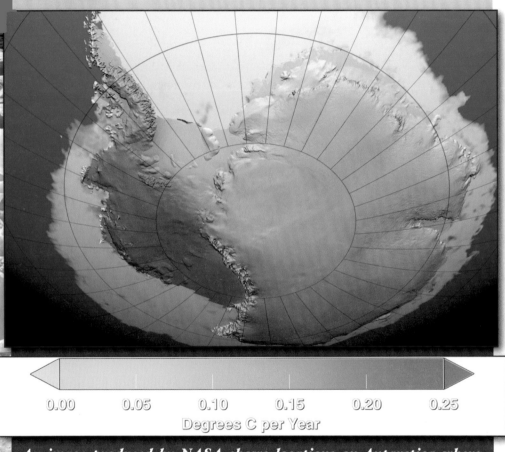

An image produced by NASA shows locations on Antarctica where temperatures increased between 1959 and 2009. The red represents areas where temperatures (measured in degrees Celsius) increased the most over the period, and the dark blue represents areas with a lesser degree of warming. **NASA/Goddard Space Flight Center Scientific Visualization Studio**

ANTARCTIC MOUNTAINS AND SURROUNDING OCEANS

The Transantarctic Mountains extend for more than 2,000 miles (3,200 kilometers) across the continent, dividing the ice sheet into two parts. The larger, eastern part rests on land that is mostly above sea level. It has been there at least 14 million years. The smaller, western part is on land that is mostly below sea level. Scientists think that if the world were to warm a little, as it has in the past, the western part could melt— perhaps in as little as a 100-year period. The melted ice would raise sea level throughout the world by about 20 feet (6 meters).

Other mountain ranges include the Prince Charles Mountains and smaller groups near the coasts. The Antarctic Peninsula has many mountains. The Ellsworth Mountains are Antarctica's highest, the Vinson Massif rising 16,066 feet (4,897 meters) above sea level. Mountains with only their peaks showing through the ice (called nunataks) are found in some areas. Several active volcanoes on the continent provide spectacular and scenic landforms at many places and are located near the Antarctic Peninsula and in the Transantarctic Mountains.

Mt. Tyree is Antarctica's second highest mountain at approximately 16,000 feet (4,877 meters). Mt. Tyree is located in the Ellsworth Mountain range. Gordon Wiltsie/National Geographic Image Collection/Getty Images

Only about 2 percent of Antarctica is free from ice. These unusual land areas, called oases, generally are near the coast and include the dry valleys of southern Victoria Land and the Bunger Oasis in Wilkes Land. High rims at the end of the valleys prevent entry of large glaciers. The warm local climate melts the ends of smaller glaciers extending into the valleys.

Surrounding Antarctica are the southern parts of the Pacific, the Atlantic, and the Indian oceans. The Antarctic Convergence, which encircles Antarctica roughly 1,000 miles (1,600 kilometers) off the coast, divides the cold southern water masses and the warmer northern waters. The Antarctic Circumpolar Current, the world's largest ocean current, moves eastward around the continent at an average speed of about half a knot (1 kilometer per hour). Sea ice up to 10 feet (3 meters) thick forms outward from the continent every winter, making a belt 300 to 1,000 miles (500 to 1,600 kilometers) wide. Even in summer the sea ice belt is 100 to 500 miles (160 to 800 kilometers) wide in most places.

THE SOUTH POLES

Antarctica has three points that are called south poles. The best known is the geographic South Pole, at 90° S. latitude on the axis of the Earth's rotation. The geomagnetic south pole is at about 78° S. 110° E., in East Antarctica; it is the center of the Southern Hemisphere auroras. The magnetic south pole is the area toward which compasses point; it is just off the Adélie Coast at about 65° S. 140° E.

Antarctica does not have 24-hour periods broken into days and nights. At the South Pole the sun rises on about September 21 and moves in a circular path upward until December 21, when it reaches about 23.5° above the horizon. Then it circles downward

This time release photo taken at the Amundsen-Scott Station, Antarctica, shows the sun moving in a horizontal line at the South Pole. Each year, the sun rises on about September 21 and moves in a circular path upward until December 21, when it reaches about 23.5° above the horizon. Then it circles downward until it sets on about March 22. Thomas J. Abercrombie/National Geographic Image Collection/Getty Images

until it sets on about March 22. This "day," or summer, is six months long. From March 22 until September 21 the South Pole is dark, and Antarctica has its long "night," or winter.

According to scientific theory, some 200 million years ago Antarctica was joined to South America, Africa, India, and Australia in a single large continent called Gondwanaland. There was no ice sheet, and trees and large animals flourished. Today, only geological formations, coal beds, and fossils remain as clues to Antarctica's warm past.

CLIMATE

Antarctica is the coldest continent. The world's record low temperature of −128.6 °F (−89.2°C) was recorded there. The mean annual temperature of the interior is −70 °F (−57 °C). The coast is warmer. Monthly mean temperatures at McMurdo Station range from −18° F (−28 °C) in August to 27 °F (−3°C) in January. Along the Antarctic Peninsula temperatures have been as high as 59° F (15 °C).

Because it is such a large area of extreme cold, Antarctica plays an important role in global atmospheric circulation. In the tropics the sun warms the air, causing it to rise and move toward the poles. When these

air masses arrive over Antarctica, they cool, become heavier, and fall from the high interior of the continent toward the sea, making some Antarctic coasts the windiest places in the world. Winds on the Adélie Coast in the winter of 1912 to 1913 averaged 40 miles (64 kilometers) per hour 64 percent of the time, and gusts of nearly 200 miles (320 kilometers) per hour have been recorded.

Antarctica's interior is one of the world's major cold deserts. Precipitation (if melted) averages only 1 to 2 inches (2.5 to 5 centimeters) a year.

PLANT AND ANIMAL LIFE

The severe climate has kept much of the Antarctic biome nearly devoid of life. Nevertheless, botanists have found bacteria and yeast growing just 183 miles (295 kilometers) from the geographic South Pole. A lichen was found in a sunny canyon 266 miles (428 kilometers) from the pole, and a blue-green alga in a frozen pond 224 miles (360 kilometers) from the pole. Microbes related to lichens colonize in green and brown layers just beneath the surface of rocks facing the sun. Mosses and liverworts grow in some ice-free areas along the coast. Two species of

These five species were found by scientists after the breakup of two ice shelves in Antarctica. They include (from top, and left to right) a new species of Epimeria; *an Antarctic male sea spider bearing its eggs; a giant Antarctic barnacle; a new giant Antarctic amphipod crustacean; and a still unidentified Antarctic sea star.* **AFP/Getty Images**

flowering plants—a grass and an herb—grow on the peninsula.

The native land animals are limited to arthropods (such as insects), which become inactive during the coldest months. Nearly all of the species are found only in Antarctica. These springtails, midges, and mites live generally along the coast among plant colonies. The southernmost known animal, the mite, has been found 315 miles (507 kilometers) from the South Pole.

The immense numbers of birds and seals that live in Antarctica are, properly speaking, sea animals. They spend most of their time in or over the water, where they get their food. These animals come ashore only to establish rookeries and breed.

About 45 species of birds live south of the Antarctic Convergence. Two penguin species—the emperor and the Adélie—are distributed widely around the entire coastline. Gentoo and chinstrap penguins occupy Antarctic Peninsula coasts and some islands. Penguins are excellent swimmers and catch and eat their food—mostly krill (a shrimplike animal) and fishes—underwater.

Four species of seals breed almost exclusively in the Antarctic. They are the Weddell seal, which ranges as far south as the sea does and can dive as deep as 2,000 feet (600 meters) for nearly an hour; the crabeater seal, which spends most of its time around pack ice (sea ice); the leopard seal, which favors penguins as its food; and the Ross seal, rarely seen. Other Antarctic species include the fur seal and the huge elephant seal. Most populous is the crabeater, whose numbers are estimated to be in the millions. Weddell seals by comparison number only 500,000. The others exist in even smaller numbers.

A leopard seal rests on an ice floe in Antarctica. **Shutterstock.com**

Fishes peculiar to the Antarctic include the Antarctic cod and the icefish. These and other Antarctic fish have developed blood that enables them to live in seawater as cold as 28° F (–2 °C).

The most important single member of the Antarctic marine food chain is the krill. This crustacean looks like a small shrimp and exists in huge numbers; one vast swarm stretching several miles in length was observed from ships, and some biologists think the total population may be 5 billion tons or more. Krill eat small marine plants (phytoplankton) and animals (zooplankton) and in turn are eaten in great numbers by squid, birds, seals, and whales.

PENGUINS

Seen from a distance, a colony of penguins might easily be taken for a group of little men. These sea birds stand erect and flat-footed and are often drawn up in long regular files like soldiers. They walk with a tread so stately and dignified that the sight is very comical. In the species known as the king penguin, the resemblance to man is heightened by the grayish-blue coat on the back. This is set off by black plumage on the head, a white breast, and a yellow "cravat" at the throat.

Prehistoric penguins stood 6 feet (1.8 meters) tall. The largest penguins today, members of the emperor penguin species, stand

Frontal view of three Adelie penguins.
Shutterstock.com

about 3 $^1/_2$ feet (1 meter) high and weigh about 80 pounds (36 kilograms).

Penguin ancestors could fly as well as any other sea bird. Now its wings are short, paddle-like flappers that are entirely useless for flight. The bird has lived for ages in or near the Antarctic regions, where it has few human or animal enemies. Thus it came to spend all of its time on land or in the water. For generations it did not fly. In the course of long evolution, its wings became small and stiff and lost their long feathers. Now they cannot be moved at the middle joint as can the wings of flying birds.

The penguins, however, became master swimmers and divers. Of all birds, they are the most fully adapted to water. Their thick coat of feathers provides a smooth surface that is impenetrable to water. Their streamlined bodies glide through the water easily. The birds use their wings as swimmers use their arms in a crawl stroke, and they steer with their feet. Penguins can swim at speeds of more than 25 miles (40 kilometers) per hour. When they want to leave the water, they can leap as much as 6 feet (1.8 meters) from the water's surface onto a rock or iceberg.

Flocks of penguins may stay at sea for weeks at a time. They resemble schools of dolphins as they leap in graceful arcs from the water to take breaths of air. Penguins feed underwater on fish, squids, and crustaceans. In fact, penguins do not know how to eat on land, and must slowly acquire this skill when captured for zoos.

In addition to Antarctica, penguins live on subantarctic islands and on cool coasts of Africa, New Zealand, Australia, and South America. Some species migrate long distances inland to ancestral nesting grounds. Because they cannot walk well on their short legs, the birds often toboggan over the ice on their stomachs.

Most penguins build a nest on the ground from pebbles, mud and vegetation, or any materials that are available. The female then lays one or two chalky white eggs. The eggs are incubated in turns by both parents, one remaining on the nest while the other goes off to feed. The breeding behavior of the emperor penguin is quite different. After laying her single egg, the female leaves to feed. The male incubates the egg by himself, cradling it between the top of his feet and his stomach. For two months the colonies of fasting males remain on the icy nesting grounds, warming and protecting their eggs in temperatures as cold as $-40°$ F ($-40°$ C). By the time the eggs have hatched, the males have lost a third of their body weight. When the females return to care for the chicks, the males are at last free to return to the water for food and rest.

CHAPTER 4
ANTARCTIC RESOURCES

I n one analysis of the potential resources that exist in Antarctica, "Antarctic natural resources" were defined as "any natural materials or characteristics (in the Antarctic region) of significance to man." By this broad definition, the term includes not only biological and mineral resources but also the land itself, water, ice, climate, and space for living and working, recreation, and storage.

CONSERVATION AND DEVELOPMENT

Antarctica is so far from world markets, and its environment is so hostile, that little economic development has taken place. Also, little is known about the amounts of natural resources that exist there. But, if world shortages of food and energy products become severe, Antarctica may be more intensely explored. In anticipation of such a need, numerous nations have signed a convention for the conservation of living marine resources on the Antarctic continent.

WHALING AND FISHING

The first people to make money by going to Antarctica were whalers and sealers. Seal hunters began catching Antarctic seals for their oil and fur in the early 1790s. Fur seals and then elephant seals were reduced almost to extinction by the mid-1800s, at which point the sealers finally stopped their Antarctic hunts. The populations of fur and elephant seals once again are growing. In 1978 the nations interested in Antarctica agreed to prohibit the taking of fur, elephant, and Ross seals. This pact also limits the annual catch of crabeater, leopard, and Weddell seals. But no seal hunting has taken place in Antarctica since 1964.

Whaling began in Antarctic waters in the 19th century. The industry enlarged greatly in the early 1900s, when steamships, harpoon guns, and shore processing stations (notably at South Georgia) were introduced. During the 1912–13 season 10,760 whales were caught. After that time nearly all the whales caught in the world were caught in Antarctic waters.

In 1931, a peak year, 40,199 whales were caught in the Antarctic, while only 1,124 were caught in the rest of the world. So many whales were caught that their numbers

Workmen dissect a whale carcass in Antarctica in 1935. Hulton Archive/Getty Images

declined, just as had those of the seals. The industry declined after 1960. In the 1980–81 season fewer than 6,000 whales were caught in the Antarctic; all were minke whales, a relatively small-sized species also called the lesser rorqual. Today these whales, as well as a few other species, are still hunted by some countries.

Commercial fishing was begun by the Soviet Union in 1967. In 1971 a Soviet fleet of 40 trawlers and support ships in the southern ocean landed an estimated 300,000 tons—mostly cod, herring, and whiting. Today fleets of other nations, mainly Japan and Norway, also fish the waters.

Krill are a plentiful food source in the frigid waters of the Antarctic. George F. Mobley/National Geographic Image Collection/Getty Images

Krill fishing began in the early 1970s and hit its peak thus far about 10 years later with an annual harvest of some 550,000 metric tons. Figures for the year 2000 put the krill intake at about 100,000 metric tons, with just a few nations participating in the industry. Though a multination fishing treaty allows larger harvests, the economic costs of catching and transporting the krill from the frigid Antarctic environment prevent most countries from taking part.

In 1982 the nations that were interested in Antarctica set up a scientific committee to study the Antarctic ecosystem and a commission to set catch limits. The nations wanted to protect the unique ecosystem and to avoid any activities that had already reduced the numbers of whales and seals in the area.

MINERALS AND OTHER RESOURCES

Minerals and petroleum have never been exploited in Antarctica. Minerals have been found in great variety but almost always in small amounts. Large mineral deposits probably exist, but the chances of finding them are small. Manganese nodules on the ocean floor, geothermal energy, coal, petroleum, and natural gas are potential resources that could

perhaps be exploited in the future. Only two large mineral deposits have been found: iron ore in the Prince Charles Mountains and coal in the Transantarctic Mountains. But it would cost too much to get these materials to market to make them economically attractive.

Some authorities think that large reserves of oil and natural gas exist in Antarctica,

Tourists in a boat watch chinstrap penguins atop an iceberg in Antarctica's waters. **DreamPictures/Photographer's Choice/Getty Images**

simply because the continental shelf is so large. However, little exploration has been done, and even if they are found, extraction will be difficult. The edge of the Antarctic continental shelf is 1,000 to 3,000 feet (300 to 900 meters) deep, much deeper than the world average continental shelf depth of about 600 feet (200 meters), and Antarctica's icebergs would threaten drill rigs. Also, the environmental impact of spills would be greater in Antarctica because low temperatures retard the growth of biological organisms that reduce crude oil to environmentally harmless components.

Some people have devised ingenious schemes for towing Antarctic icebergs north to warm, dry lands as a cheap source of fresh water. But many scientists and engineers believe that an iceberg, even if protected by the best possible means, would break and melt before it arrived at the place where it would be used.

Commercial tourist visits to Antarctica began in the 1950s. Between 1958 and 1980 an estimated 16,640 passengers on 80 ship cruises visited places along the Antarctic Peninsula and in the Ross Sea. In contrast, during the 2005–06 summer season alone,

26,245 tourists, mostly onboard commercial vessels, visited the Antarctic continent.

POLITICAL AND INTERNATIONAL RELATIONS

Because it has never had permanent human settlements, Antarctica has had an unusual political history. Seven nations have claimed pie-shaped sectors of territory centering on the South Pole. Three of the claimed sectors overlap on the Antarctic Peninsula. One sector is unclaimed. Most other nations do not recognize these claims. The United States policy, for example, is that the mere discovery of lands does not support a valid claim unless the discovery is followed by actual settlement. Also, like many other nations, the United States reserves all rights resulting from its explorations and discoveries.

This unsettled situation might have continued had it not been for a surge of scientific interest in Antarctica that developed in the mid-1950s. At that time scientists of 12 nations decided to make research in Antarctica a major portion of the IGY. The 12 nations were Argentina, Australia, Belgium, Chile, France, Japan, New Zealand,

In April 1958, during the International Geophysical Year (IGY), meteorologists at the Amundsen-Scott Station, Antarctica, study polar weather's global impact. Thomas J. Abercrombie/National Geographic Image Collection/Getty Images

Norway, South Africa, the United Kingdom, the United States, and the Soviet Union. When the IGY was completed in 1958, these nations decided to continue their research programs in Antarctica.

Much of the research had been achieved through international cooperation, and the 12

nations carried their new, friendly ties from science into politics. They met in Washington, D.C., in 1959 to write the Antarctic Treaty. The treaty reserves the region for peaceful purposes, especially scientific research. It prohibits nuclear weapons and disposal of radioactive waste, and it does not allow military activities except to support science and other peaceful pursuits. The treaty does not recognize or dispute the territorial claims of any nation, but it also does not allow any new claims to be made. It allows members to inspect each others' installations, encourages the exchange of personnel, and requires each nation to report to the others on its plans and results.

The treaty does not include anything about sharing Antarctica's natural resources, but it does provide for meetings every other year to further its objectives. At these meetings the treaty nations have agreed on conservation plans and on responsible collection and sharing of resources. Other nations later joined the Antarctic Treaty, and by 2006 a total of 46 countries had signed it.

Most early Antarctic expeditions were directly or indirectly of economic incentive. For some, it was the search for new trading routes; for others, it meant the opening of new fur-sealing grounds; still others saw a possibility of mineral riches. Although early explorations were nationalistic, leading to territorial claims, modern ones have come under the international aegis of the Antarctic Treaty and are scientific in nature, helping lead to a better understanding of the total world environment.

SCIENTIFIC RESEARCH

Every year about 25 nations send scientists to Antarctica to do research. In the Antarctic summer about four thousand people are in the region for this work. They operate research stations and camps; travel in airplanes, helicopters, and snowmobiles to the areas that they need to study; and operate

Scientists measure snow and ice thickness on the frozen Bellingshausen Sea, Antarctica. **Maria Stenzel/National Geographic Image Collection/ Getty Images**

ships for resupply and oceanic research. In winter about one thousand people remain to operate more than 35 research stations scattered around the continent. The winter inhabitants are isolated for several months at a time because it is too cold for anyone to get to them, even in airplanes. Biologists, geologists, oceanographers, geophysicists, astronomers, glaciologists, and meteorologists conduct experiments here that cannot

be duplicated anywhere else. In the 1970s researchers began taking measurements of the protective ozone layer in the atmosphere over Antarctica. By the mid-1980s scientists discovered that a "hole" developed periodically over the region; it was found that the ozone layer there was thinned by as much as 40–50 percent from its normal concentrations. This severe regional ozone depletion was explained as a natural phenomenon, but one that was probably increased by the effects of chemicals called chlorofluorocarbons (CFCs) that were used in refrigerators, air conditioners, and spray cans. Concern over increasing global ozone depletion led to international restrictions on the use of CFCs and to scheduled reductions in their manufacture. The Antarctic ozone hole grew in size throughout the 1990s and the first decade of the 21st century, but despite these findings, most scientists contend that the ozone layer will eventually recover. Signs of recovery might not become apparent until about 2020, however, because of natural variability.

EARLY EXPEDITIONS

The first expedition to come close to Antarctica took place from 1772 through

Captain James Cook. **Time & Life Pictures/Getty Images**

1775. The English navigator James Cook sailed around the continent and came within 100 miles (160 kilometers) of it. Land was seen in about 1820, when British and United States seal hunters and a Russian exploring expedition reached the Antarctic Peninsula.

In the Antarctic summer of 1839–40 a United States Navy expedition headed by Charles Wilkes mapped 1,500 miles (2,400 kilometers) along the coast of East Antarctica. The next summer James Clark Ross of Great Britain sailed into the Ross Sea, traveling as far south as a ship can go. The first recorded landing on Antarctica was on Cape Adare in 1895, and the first group to spend a winter there did so from March 1898 to March 1899.

Captain Robert F. Scott, equipped for his last journey to the South Pole. **Hulton Archive/Getty Images**

The struggle inland and toward the geographic South Pole began with the first expedition by Robert F. Scott of Great Britain in 1901–04. But the first person to reach the pole was Roald Amundsen of Norway on Dec. 14, 1911. On another Antarctic expedition Scott arrived at the pole just a month later; he died on March 29, 1912, trying to return to the coast.

LATER EXPEDITIONS AND STUDIES

Early expeditions to Antarctica relied on sail power, dog power, and human power for their transportation. The mechanical age arrived on Nov. 16, 1928, when George Hubert Wilkins, leading an American expedition, made an airplane flight from Deception Island. On Nov. 29, 1929, Richard E. Byrd of the United States flew a three-motor Ford plane over the South Pole. Byrd also explored parts of Antarctica by air and on the surface in 1933–35 and 1939–41 and commanded the largest single expedition ever made to Antarctica—the United States Navy's Operation High Jump in 1946–47. Thirteen ships, many airplanes and helicopters, and

Members of Will Steger's International TransAntartic expedition team at the Admundsen-Scott Station at the South Pole a few weeks after they arrived on December 11, 1989. **AFP/Getty Images**

thousands of men made surveys almost all the way around the continent.

In 1990 a six-man international expedition led by an American named Will Steger completed a 221-day trek across Antarctica from west to east using dogsleds. At more than 3,700 miles (6,000 kilometers), it was the longest dogsled trek, as well as the first unmechanized passage through the South Pole. The team members were from the United States, the Soviet Union, France, China, Japan, and Great Britain.

In 2008 an expedition of American and Australian scientists found evidence in the Transantarctic Mountains that supported the geologic theory that East Antarctica was connected to the western edge of North America 600 million to 800 million years ago.

Members of the international Kaspersky Commonwealth Antarctic Expedition sing and dance on January 2, 2010, to celebrate the new year during their expedition to the South Pole. **AFP/Getty Images**

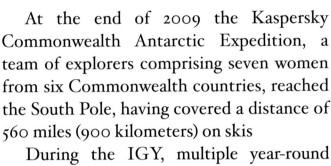

At the end of 2009 the Kaspersky Commonwealth Antarctic Expedition, a team of explorers comprising seven women from six Commonwealth countries, reached the South Pole, having covered a distance of 560 miles (900 kilometers) on skis

During the IGY, multiple year-round stations were established on Antarctica, including one at the geographic South Pole and one at the south geomagnetic pole. Since then scientific studies have been carried out continuously by numerous nations. In 1991 the voting members of the Antarctic Treaty developed a protocol on environmental protection that forbids commercial mining in Antarctica for the next 50 years. The protocol was officially adopted in 1998.

CONCLUSION

For hundreds of years the icy areas at each end of the Earth have challenged explorers, but in recent decades the polar biomes have become increasingly accessible. In the Arctic, the discovery phase of exploration is over; there is no longer any possibility of finding new lands. Photo surveys have provided accurate maps, and improved aircraft and base facilities have been established. In Antarctica, the most remote and formidable continent, all of the mountain regions have been mapped and visited by geologists, geophysicists, glaciologists, and biologists, and many hidden ranges and peaks are known from geophysical soundings of the Antarctic ice sheets. In both polar regions, much of the emphasis now is on scientific investigations that reflect pressing environmental concerns. A considerable amount of research continues to be focused on the relationship between the Arctic regions and global warming and climate change, while in Antarctica, researchers work to address many problems

of global consequence, including taking measurements and conducting studies of the stratosphere's apparently endangered ozone layer. Over the last half century, internationally supported programs have facilitated incredible advances in polar science, and cooperation among nations will remain of vital importance to future scientific efforts as well as to the responsible stewardship of these vast biomes.

biome The largest geographic biotic unit, a
major community of plants and animals
with similar life forms and environmental
conditions.

calving The separation of a volume of ice
from its parent glacier.

firn The partially compacted granular snow
that forms the surface part of the upper
end of a glacier.

floe Floating ice formed in a large sheet on
the surface of a body of water.

gale Wind that is stronger than a breeze;
specifically a wind of 28–55 knots (50–102
kilometers per hour).

geothermal Of, relating to, or using the
heat of the Earth's interior.

Gondwanaland Name given by geologist
Edward Suess to an ancient super-
continent comprised of modern-day South
America, Africa, Arabia, Madagascar,
India, Australia, and Antarctica.

krill Small, shrimplike crustaceans that
are an important source of food for fish,
squid, whales, seabirds, and other animals,
especially around Antarctica. They swim
in large swarms, and can grow to be about
2.5 inches (6 centimeters) long.

massif A mass of compact and largely
isolated mountains.

microbes Germs or microorganisms.

mites Tiny animals, related to spiders, that often infest animals, plants, and stored foods.

permafrost A permanently frozen layer at variable depth below the surface in frigid regions of a planet (as Earth).

rookery Breeding ground or haunt especially of social birds or mammals (as penguins or seals), or a colony of such birds or mammals.

sediment Matter, such as dirt or rocks, that is deposited by water.

sod The upper layer of soil bound by grass and plant roots into a thick mat.

tundra An ecosystem defined by great expanses of treeless ground and a harsh, frigid climate.

Arctic Institute of North America
University of Calgary
2500 University Drive N.W.
Calgary, AB T2N 1N4
Canada
(403) 220-7515
Web site: http://www.arctic.ucalgary.ca
The institute's mandate is to advance the
    study of the North American and circum-
    polar Arctic through the natural and social
    sciences, and the arts and humanities,
    and to acquire, preserve, and disseminate
    information on physical, environmental,
    and social conditions in the north.

Canadian Ice Service (CIS)
Environment Canada
Inquiry Centre
351 St. Joseph Boulevard.
8th Floor, Place Vincent Massey
Gatineau, QC K1A 0H3
Canada
(800)668-6767
Web site: http://www.ec.gc.ca/glaces-ice
The mission of the Canadian Ice Service
    (CIS) is to provide the most accurate and
    timely information about ice in Canada's
    navigable waters. The CIS works to

promote safe and efficient maritime operations and to help protect Canada's environment. The Web site contains background about ice climatology, educational resources, and more about sea ice and icebergs.

National Snow and Ice Data Center
CIRES, 449 UCB
University of Colorado
Boulder, CO 80309
(303) 492-2468
Web site: http://nsidc.org
The National Snow and Ice Data Center offers information on the various aspects of snow and ice and how scientists set about researching the poles.

United States Antarctic Program
National Scientific Foundation
Office of Polar Programs
4201 Wilson Boulevard, Suite 755
Arlington, VA 22230
(703) 292-5111
Web site: http://www.usap.gov
Overview of the science of U.S. explorations in the Antarctic and scientific discoveries, including Web cams of various Antarctic stations and the *Antarctic Sun*, a

newspaper about the ice, the people, and the program.

World Wildlife Fund (WWF)
1250 Twenty-Fourth Street N.W.
P.O. Box 97180
Washington, DC 20090-7180
(202) 293-4800
Web site: http://www.worldwildlife.org/
    what/wherewework/arctic/index.html
The World Wildlife Fund works to protect
    the delicate ecosystem in the Antarctic
    against such threats as global warming,
    species depletion, and overfishing.

WEB SITES

Due to the changing nature of Internet links, Rosen Educational Services has developed an online list of Web sites related to the subject of this book. This site is updated regularly. Please use this link to access the list:

http://www.rosenlinks.com/ies/polr

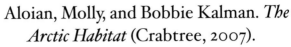

Aloian, Molly, and Bobbie Kalman. *The Arctic Habitat* (Crabtree, 2007).

Dewey, Jennifer. *Antarctic Journal: Four Months at the Bottom of the World* (HarperCollins, 2001).

Gogerly, Liz. *Amundsen and Scott's Race to the South Pole* (Heinemann, 2007).

Kalman, Bobbie, and Rebecca Sjonger. *Explore Antarctica* (Crabtree, 2007).

Lynch, Wayne, and Aubrey Lang. *The Arctic* (NorthWord, 2007).

Myers, W.D. *Antarctica: Journeys to the South Pole* (Scholastic, 2004).

Roberson, Dennis. *Antarctica* (Lucent, 2003).

Roza, Greg. *An Arctic Ecosystem* (PowerKids Press, 2009).

Whitehouse, Patricia. *Living in the Arctic* (Rourke, 2007).

Woods, Michael. *Science on Ice: Research in the Antarctic* (Millbrook, 1995).

A

Amundsen, Roald,
30–33, 75
Andrée, Salomon, 34–35
Antarctica
climate of, 17, 51–52
conservation and
development of,
60–67
exploration of, 66, 70,
72–78
geography of, 43–59
plant and animal life,
52–59
political and interna-
tional relations, 67–69
resources of, 60–69
scientific research and,
42, 43, 67–69, 70–72,
77, 78
tourism to, 43, 66–67
Antarctic Treaty, 69, 70, 78
Arctic Circle, 11–12, 17
Arctic Ocean, 12, 17, 19,
20, 22, 25, 33
Arctic region
area of, 11–12
climate of, 16–19
environmental threats
to, 20
exploration of, 28–30,
31–36, 79
geography of, 10–27
land of, 13–15
people of, 24–25, 26
plant and animal life,
19–22
scientific research and,
10, 20, 28–29, 38–42
Arctic Research
Laboratory Ice Station
II (ARLIS II), 39

B

Byrd, Richard E., 75

C

Cook, F. A., 36
Cook, James, 74

F

fishing, 24, 25, 63–64
Franklin, Sir John, 28–29, 31

G

global warming, 16, 19, 20,
24, 42, 79
Greely, A. W., 29

I

Intergovernmental Panel
on Climate Change
(IPCC), 19
International Geophysical
Year (IGY), 38–39,
67–68, 78

International Polar Year, 42
Inuit, 10, 23, 24–25

K

Kaspersky
    Commonwealth
    Antarctic
    Expedition, 78

L

Lapland, 25–27

M

McClure, Robert, 29–30
MacMillan, Donald B., 36
minerals, 40, 41–42, 64–65

N

Nansen, Fridtjof, 32, 33
natural gas, 40, 42, 64–66
Nobile, Umberto, 32–33
Nordenskjöld, Nils Adolf
    Erik, 30
North Pole, 12, 17, 28, 29,
    30, 31, 32, 39, 42
Northwest Passage, 28,
    30, 32

O

oil/petroleum, 20, 25,
    40–41, 42, 64–66

P

Peary, Robert E., 32, 35–36
penguins, 54, 55, 56–59
polar bears, 21, 22–24

R

reindeer, 21, 26, 27
Ross, James Clark, 74

S

Sami (Lapps), 25, 26–27
Scott, Robert F., 32, 75
seals, 54–55, 61, 62, 64,
    70, 74
South Pole, 30, 32, 43,
    49–51, 52, 54, 67, 75,
    77, 78
Stefansson,
    Vilhjalmur, 36
Steger, Will, 76–77

T

tundra, 19–21, 25, 26

U

U.S. Navy, 39, 74, 75

W

whaling, 61–63, 64
Wilkes, Charles, 74
Wilkins, George Hubert, 75